# Gardens
## ON A GRAND SCALE

*Bowood*

# CREATING THE

One fine day in 1529 Henry VIII had a very good idea – to construct a garden at Hampton Court rivalling the gardens of the French king. Henry's garden was huge and dazzling with large squares of grass, coloured brickdust and sand, heraldic beasts on poles and wondrous topiary. On the riverbank stood a banqueting house atop a large mound, and a watergate received important guests. This was England's prototype garden on a grand scale.

The English remained obsessed by European garden design, particularly the stylised creations of the Frenchman André le Nôtre, whose vast rectangular ponds, spectacular fountains and elaborate parterres were copiously copied. Hampton Court and parts of Chatsworth are fine examples of French

influence, whilst Westbury Court in Gloucestershire is a unique example of a seventeenth-century Dutch water garden, and the remarkable topiary at Levens Hall in Cumbria was also created at that time in the Dutch style.

All changed dramatically in the mid-eighteenth century with the arrival of Lancelot Brown, nicknamed 'Capability' because of his talent for assessing 'the capabilities of nature' to enhance what was already there. By sweeping away the geometric formality of gardens modelled on Versailles, he invented the landscaped garden. William Kent and Charles Bridgeman had previously paved the way, with Kent trying to introduce the romantic images of Claude Lorrain's

*Hampton Court by Leonard Knyff*

# GRAND DESIGN

landscape paintings into the garden, and Bridgeman introducing the ha-ha which integrated garden and parkland. They provided a link with the formality of previous gardens but lacked the vision, energy and panache of Brown who fashioned arcadian landscapes of majestic lakes, undulating slopes and distinctive clumps of trees. It was this magical formula which captivated the autocratic eighteenth-century aristocracy, producing a profound and lasting impact on the English countryside.

Blenheim, Broadlands, Bowood, Longleat, Stowe, Chatsworth, Claremont, Leeds Castle, Warwick Castle, Syon and Harewood are all Brown's work. In this golden age of English garden design the nation's leaders were wealthy, self-indulgent and totally free from planning restraints. At Chatsworth the Duke of Devonshire moved the village of Edensor and widened the River Derwent, whilst Lord Milton demolished a hundred homes to improve his park at Milton Abbas in Dorset. Not surprisingly, 'Capability' Brown, the man born of a flirtation between a north-country squire and a chambermaid, eventually became the king's head gardener and high sheriff of Huntingdon, acquiring all the wealth and status bestowed by such an office.

Meanwhile, Henry Hoare, wealthy banker and remarkable amateur gardener, created Stourhead, the enchanting English water garden in Wiltshire.

*Hestercombe, Somerset -
the inspired creation of Gertrude Jekyll
and Edwin Lutyens*

*Lancelot
'Capability'
Brown*

*Heligan, Cornwall*

The most influential designer to follow Brown was Humphry Repton, who laid out Woburn in 1802 and also extended Brown's work at Sheffield Park in East Sussex, another stunning water garden. Repton reintroduced a degree of formality into gardens, with terraces and flights of steps.

Towards the end of the nineteenth century, a chance encounter led to the unlikely but brilliant partnership between Edwin Lutyens and Gertrude Jekyll. The young architect was half the age of Jekyll, but together they

*Gertrude Jekyll*

*Sculpture at Roche Court in Wiltshire*

*Coleton Fishacre, Devon*

*Stonework at Hestercombe, Somerset*

## Glossary

**Arboretum** *a garden devoted to specialist trees*

**Ha-ha** *a ditch with a wall at the boundary of a garden or park, allowing uninterrupted views of the countryside beyond*

**Parterre** *a level ornamental garden, with paths between the formally arranged flowerbeds*

**Pinetum** *a plantation of pine trees*

**Rill** *a small stream or water channel*

**Topiary** *the practice of clipping trees or shrubs into decorative shapes*

created more than seventy gardens, Hestercombe in Somerset being a marvellous example. Lutyens provided the architectural framework for Jekyll's superb use of colour in her planting. Originally an artist, her vision was that of the Impressionists, and she elevated the flowerbed to an art form. Jekyll's work is also at nearby Barrington Court.

Geoffrey Jellicoe, Harold Peto and Norah Lindsay took the tradition further into the twentieth century – Peto's work can be seen at Buscot in Oxfordshire, another exquisite water garden – whilst Vita Sackville-West created Sissinghurst, perhaps one of the greatest gardens in the world, in the 1930s. Another talented amateur, Lady D'Oyly Carte, carved out a stunning seaside garden at Coleton Fishacre in south Devon.

The West Country also pioneered the perfect contemporary garden as Barbara Hepworth's sculptures spilled out of the studio into her garden at St Ives near the tip of Cornwall, thereby introducing the sculpture garden, a concept excellently developed at Roche Court in Wiltshire and further north at Bretton Hall's Yorkshire Sculpture Park, which features the work of Henry Moore.

Precisely how Henry VIII would have rated the work of Barbara Hepworth and Henry Moore against his heraldic figures at Hampton Court remains a matter of conjecture, but for nearly 500 years gardens have been created on a grand scale all over England – and they continue to provide enormous pleasure to their millions of visitors.

*Overbecks, Devon*

### The Emperor Fountain

In 1847 the sixth duke learnt that Czar Nicholas, Emperor of Russia, was visiting England and might be coming to Chatsworth. He decided to greet him with an even larger fountain than that at the Czar's Russian Imperial Palace, and set Joseph Paxton to work on this project. A vast tonnage of earth was dug out to form an eight-acre lake to power the fountain, and such was the urgency of the work, that digging continued at night using flares! It was finished in six months. The Czar never visited Chatsworth but this splendid fountain, capable of jetting water a hundred metres into the sky, remains to this day.

### The view across the West Garden

Beyond is the landscaped park with the River Derwent winding through, a landscape little changed since 'Capability' Brown first laid it out in the mid-eighteenth century.

# CHATSWORTH – THE SECOND VERSAILLES

Boasting a garden of over one hundred acres and a landscaped park of more than a thousand, Chatsworth can truly be described as a garden on a grand scale. Indeed, so generous were its proportions, that visitors to the house in the mid-nineteenth century who wished to see the kitchen garden were advised to call a carriage!

Home of the Duke of Devonshire, this magnificent house is superbly sited on the banks of the River Derwent against a backdrop of heavily wooded hills. Its garden has reflected the changing fashions of successive centuries in such stunning style that Charles de Saint Amant, the nineteenth-century governor of the Tuileries in Paris, even felt compelled to write a book on Chatsworth, describing it as 'the second Versailles'.

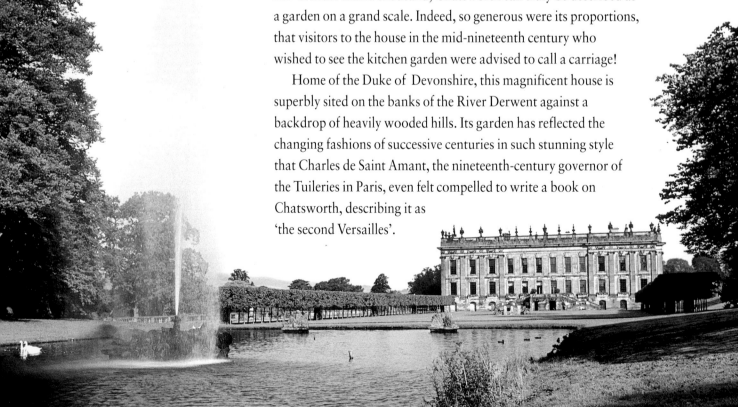

# Harewood House
## GREAT GARDEN OF THE NORTH

The Earl of Harewood's park retains the look of 'Capability' Brown's work in the 1770s, contrasting with Sir Charles Barry's more formal terraces of a century later, with intricate parterre, fountains and extensive herbaceous borders. Visit in early summer when the rhododendrons are in flower around the lake.

### The Great Cascade

The first duke employed Grillet, pupil of the legendary Frenchman le Nôtre, to design the Great Cascade. Fed by streams from the moors, the water falls continuously over sixty steps in twenty-four flights from the Cascade House.

### The Conservative Wall

At first glance this wall appears to be post-modernist in style, but it was actually built in 1824 by Joseph Paxton. Paxton was only twenty-three when he was first employed as head gardener. Inside the conservatory are figs, peaches, nectarines and apricots, plus two massive flowering *Camellia reticulata* shrubs planted around 1850.

# BOWOOD – 'TRUE ART IS NATURE TO ADVANTAGE DRESS'D'
### ALEXANDER POPE

If 'Capability' Brown could return to the Earl and Countess of Shelburne's home in Wiltshire, he would not be disappointed. The view from the house across the lake to the Doric Temple in front of a distinctive clump of trees remains classic 'Capability' Brown countryside as he conceived it for the second earl nearly 250 years ago. A perfect time to see it is in the quiet of a late summer afternoon when the tall trees of the Great Park cast long shadows across sunlit lawns and are reflected in the calm waters of Brown's lake. Yet there was a price to pay for the creation of this perfect arcadian world, for a village had to be submerged when the valley was flooded to form the lake. The villagers were rehoused in the new model village of Sandy Lane a mile away, refugees of a romantic age.

**Standing guard**
A rather disgruntled statue of a lion guards the terraces.

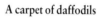

**A carpet of daffodils**
The park is planted with thousands of different varieties of daffodils, whose bold colours stand out against blue skies.

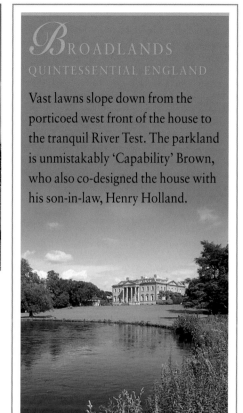

## BROADLANDS
### QUINTESSENTIAL ENGLAND

Vast lawns slope down from the porticoed west front of the house to the tranquil River Test. The parkland is unmistakably 'Capability' Brown, who also co-designed the house with his son-in-law, Henry Holland.

**A glorious frontage**

Summer rose beds and large urns of geraniums adorn the terraces in front of the house at Bowood.

**Autumnal colours** *(left)*

Autumn is the ideal time to visit Bowood. Stride out to House Hollow, Archery Lawn and the pinetum, and see for yourself the giant redwoods and cedars planted centuries ago, their leaves ablaze with colour.

## BLENHEIM 'THE FINEST VIEW IN ENGLAND'
### LORD RANDOLPH CHURCHILL

'Capability' Brown's gracious parkland provides an appropriate setting for Vanbrugh's magnificent house at Blenheim, the birthplace of Sir Winston Churchill. The master gardener refused to be intimidated by Vanbrugh, half-drowning the illustrious architect's grand bridge in the act of creating the lake, which was described by Sir Sacheverell Sitwell as having no equal: 'Nothing finer in Europe'.

# HESTERCOMBE

## 'CLASSICAL GARDEN DESIGN ON A GRAND SCALE'
### CHRISTOPHER HUSSEY

Gazing down over the vale of Taunton towards the Blackdown Hills, your eye alights on a garden that represents the pinnacle of the long and fruitful collaboration between Edwin Lutyens and Gertrude Jekyll. Situated on a sloping site to the south of an unprepossessing house, the garden consists of three terraces descending as giant steps towards open countryside. The third terrace surrounding the great 'plat' is a unique three-dimensional garden with an endless variety of intriguing aspects. Lutyens's honey-coloured ham stone garden structures and flagstone paths provide a solid framework for Jekyll's exuberant planting.

### Classical lines

Water is an important feature of the garden at Hestercombe, as witnessed in the long planted rills and the pools set in alcoves with mask water spouts.

### The rotunda

Commissioned in 1904 by Viscount Portman, Lutyens was truly inspired in his vision of Hestercombe. He cleverly linked the Dutch garden, which protrudes at an odd angle from the main garden, with the rotunda, at the same time creating an attractive feature.

### The orangery

Lutyens's Italianate orangery, built in the 'Wrenaissance' style, and the little Dutch garden were originally built on top of an old rubbish dump.

**Carved in stone**

The architectural detailing of Lutyens's stonework is superb, as can be seen on this stone mask water spout.

# Barrington Court - A Gentle Garden

Gertrude Jekyll was also involved in the planting at Barrington Court, one of a number of delightful honey-coloured Tudor manor houses near Yeovil in Somerset. It is a garden to wander around slowly on a warm summer's day, to savour gently not only its colours and forms, but also its scents, for as Jekyll's eyes faded, her planting depended increasingly on her sense of smell. Drift amongst the hollyhocks, irises, azaleas, marigolds and dahlias in the lily garden, then visit the working kitchen garden and the cider orchard. There's no need to hurry.

# Montacute

## Border Country

In front of an imposing Elizabethan house is a large smooth square of lawn edged by dazzlingly bright herbaceous borders that are surrounded by honey stone walls. The whole effect is completed by two stylish Tudor pavilions standing at the corners. Beyond is the substantial parkland that provides a relaxed venue for the annual horse trials which attract many of the country's top riders.

Vita Sackville-West, Graham Stuart Thomas and the renowned Mrs Phyllis Reiss, then owner of nearby Tintinhull, all had a part in fashioning the borders which are both a triumph of planting and a joy to behold, particularly in midsummer.

# M OTTISFONT ABBEY – THE SWEET SMELL OF SUCCESS

A nyone remotely interested in roses should beat a speedy path to this twelfth-century Augustinian priory that derives its name from the clear-flowing font or spring nearby. Set in tranquil parkland beside a tributary of the River Test, the grounds are planted with fine trees, including immense London plane trees, mature walnut, Spanish chestnut and cedars. Traces of the work of Geoffrey Jellicoe and Norah Lindsay can be found in the formal gardens near the house.

Star attraction of Mottisfont, however, is the large walled rose garden, which displays pre-twentieth-century roses from all over the world. The box-edged beds radiating from a central fountain pool contain more than 300 varieties.

**Roses for an Empress**
Many of the roses were originally developed at Malmaison for Napoleon's first wife.

**The Graham Thomas rose**
This rose was named after the creator of the rose gardens at Mottisfont.

## HYDE HALL – TRIUMPH OVER ADVERSITY

Forty years ago this beautiful garden, now owned by the Royal Horticultural Society, consisted of six trees atop a windswept hill. Its development has been an arduous task.

Early attempts to grow rhododendrons in exposed and unfavourable soil failed miserably, and roses planted in the early days deteriorated as trees matured. Today, Hyde Hall is a reminder of just what can be achieved with skill and determination.

**A scented path**
Imaginative planting along the path to the second rose garden at Mottisfont sets off the roses and provides interest throughout the season.

## SISSINGHURST – SUPERB SACKVILLE-WEST

In contemplating a group of great gardens it would be a travesty not to mention this famous creation of the late Vita Sackville-West and her husband Sir Harold Nicolson, who transformed a run-down group of buildings and land into a magical series of outdoor rooms. Their creation introduced new artistic concepts into a time-honoured craft.

Sometimes glowing tributes can raise expectations impossibly high, but Sissinghurst never fails to enchant its visitors, whatever the time of year.

# ꟻTOURHEAD 'ONE OF THE MOST PICTURESQUE SCENES IN THE WORLD'

HORACE WALPOLE

Surprisingly, this classic water garden was the work of a gifted amateur, the wealthy banker Henry Hoare. Inspired by the romantic paintings of Claude Lorrain, he spent twenty-five years creating a series of artfully placed arcadian images around a large oval-shaped lake. In so doing, he successfully upstaged the great professional landscape gardeners of the day.

Stourhead looks magnificent at any time of year. In spring the shores of the lake are yellow with daffodils, and the path that circumnavigates the lake is fringed with enormous magnolias and camellias. In summer the estate is ablaze with rhododendrons and azaleas, while in autumn it rivals New England for the spectacular colour of its trees and shrubs. Winter brings a certain brooding Wagnerian splendour, especially on a clear day *(as above)*, when everything seems to be in a state of suspended animation.

**A scene of exquisite beauty** *(right)*

The Pantheon viewed across the bridge and lake, with giant gunnera in the foreground, makes a delightful setting. Temples were not merely ornamental, they provided a changing room for bathing, and a splendid setting for picnics.

**The Palladian bridge**
'It is simple and plain,'
said Henry Hoare,
'I took it from
Palladio's bridge in
Vicenza.'

**The Temple of
Apollo**
A view of the
beautifully
proportioned work of
Henry Flitcroft, who
conceived most of the
buildings in the
gardens.

## STUDLEY ROYAL – THE TAMING OF NATURE

When John Aislabie's political career was ruined by the
South Sea Bubble scandal in 1720, he devoted the rest of his
life to creating this exceptionally fine formal water garden
in North Yorkshire, representing the fashionable
eighteenth-century idea of beauty only being successfully
attained when nature was 'tamed'. The lakes, avenues,
temples and cascades collectively conjure up an evocative
atmosphere of peace and tranquillity, particularly the
Moon Pond and Temple of Piety. The dramatic ruins of the
twelfth-century Fountains Abbey provide a focal point to
the garden, and the whole area is classified as a World
Heritage Site.

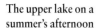
**The upper lake on a summer's afternoon**

Rhododendrons cascade down to the water's edge to meet rafts of white water lilies floating on the surface.

**Early evening at the second lake in autumn**

Trees and shrubs crowd the banks at the water's edge. 'Such is the power of vegetation at Sheffield Park that every berry soon becomes a bush and every bush a tree,' lamented Humphry Repton.

## SHEFFIELD PARK – SWAN LAKES

'Capability' Brown produced the initial two lakes for the first Earl of Sheffield, then Humphry Repton went on to create further lakes in the valley that slopes gently down from the house. The third earl's contribution was cricket and an arboretum. Early in the twentieth century Arthur Gilstrap Soames planted clumps of rhododendrons and fiery Japanese maples. Was this the horticultural equivalent of too many cooks? Not at all, for today the hundred-acre woodland garden extending around the quintet of lakes is a gracefully harmonious vista, which looks particularly beautiful in early summer and autumn.

Sheffield Park is currently owned by the National Trust, and ranks as one of the finest water gardens in the country.

**Acers in autumn**

Note the brilliant foliage of the maple trees at this time of year.

**The third lake in autumn**

Fiery Japanese maples and tall pampas grass are reflected in the calm water of the third lake at Sheffield Park.

## CASTLE HOWARD
### A GREAT NORTHERN GARDEN

Both house and grounds were designed by Sir John Vanbrugh, with some extremely impressive fountains added by Nesfield in 1853, the massive Atlas Fountain facing the house being particularly noteworthy. Outstanding features include the Great Lake covering more than seventy acres, the Temple of the Four Winds, a newly extended arboretum and one of the largest collections of old-fashioned roses in Europe.

## BUSCOT PARK
### THIS GREEN AND PLEASANT LAND

Early this century Harold Peto was commissioned to create a water garden on this extensive estate astride the River Thames above Oxford. The resulting chain of cascades, basins and canals fed by a wondrous dolphin fountain is extremely elegant and creates a scene of cool, green tranquillity. The water flows languidly through dense woods and runs under a balustraded bridge towards a large lake with a classical temple carefully placed on the far shore.

# COLETON FISHACRE
## GREENHOUSE WITHOUT GLASS

Warmed by the Gulf Stream, the south coast of Devon and
Cornwall nurtures a number of exquisite seaside gardens.
Among them is the eighteen-acre sub-tropical garden of Coleton
Fishacre, which lies in a deep combe running steeply down to the
mouth of the River Dart.

Sir Rupert and Lady Dorothy D'Oyly Carte acquired the virgin
site in 1925 to create the long, low, Lutyens-inspired house and
garden. The garden is full of rare and exotic trees and shrubs, not
often found outdoors in England. Rhododendrons, mimosas,
magnolias, camellias, dogwood and Chilean jasmine bloom here in
spring and early summer, while scarlet Japanese maples
hold forth in autumn. The whole effect is quite magical.

**The path to the sea**
A broad path winds down
through the sloping garden.

**Echoes of Lutyens**

The terrace, walls, rills and ponds are
all created in the style of Sir Edwin
Lutyens. This is not surprising, as the
house was designed by Oswald
Milne, a disciple of Lutyens, who also
shared his precise eye for detail.

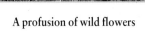

**A profusion of wild flowers**

The lower garden in spring
offers a wonderful mixture of
bluebells and wild garlic, set
against the vivid green of new
foliage and long grass.

**Trickling streams and shady pools**

Plants thrive here where the atmosphere is warm
and humid.

# $O$VERBECKS
### STUNNING GARDEN,
### SUPERB VIEWS

Another sub-tropical gem
on the south Devon coast:
this steeply terraced
garden of banana palms,
myrtle and fuchsias
overlooking an azure-blue
sea creates a gloriously
Mediterranean feel. Here
you will find a ninety-year-
old magnolia, twelve
metres high, and an
extremely rare tree,
*Cinnamomum camphora.*

*Camellia japonica*

This fine specimen comes from the national collection at Mount Edgcumbe.

**The view across Plymouth Sound from Mount Edgcumbe**

Offshore is Drake's Island, named after the famous sixteenth-century navigator.

# MOUNT EDGCUMBE

'A MOST BEAUTIFUL PLACE AS WAS EVER SEEN'    SAMUEL PEPYS (1683)

Mount Edgcumbe lies at the entrance to the Tamar estuary on land originally acquired by the Edgcumbe family in the fourteenth century for two farthings. Today the imposingly turreted mansion and estate are jointly owned by Plymouth City Council and Cornwall County Council. The estate covers more than 850 acres with ten miles of coastline around Rame Head to the open sea. Deer roam freely in the park amongst temples, statues and summer houses.

Near the house is a ten-acre formal garden protected by a twelve-metre hedge. In the Italian garden is a Doric orangery, where oranges have grown since they were brought back from Constantinople in 1744. An outstanding attraction is the National Camellia Collection blooming from January to May in the amphitheatre.

# St Michael's Mount
## A GARDEN AGAINST THE ELEMENTS

Asked to name his greatest achievement on the Mount, Lord St Levan's robust reply was, 'Bringing one hundred tons of manure over from the mainland!' The castle, on a small offshore island, is internationally famous; now the gardens are gaining a reputation in their own right, in spite of rocky soil, salt and an exposed position.

**The English garden house**
This house is just one of the many listed buildings at Mount Edgcumbe.

**Abbey ruins**
Part of the old abbey walls have been incorporated into the garden.

# Abbey Gardens,
## TRESCO, ISLES OF SCILLY

A mere two miles long, and one mile across, the island of Tresco is surrounded by some of the whitest beaches and brightest blue water anywhere in the world – a dazzling landscape scrubbed and polished by wind and tide. Masses of gorse and wild flowers make virtually the whole island a garden, the heart of which is the seventeen-acre Abbey Gardens beside the Great Pool, a large freshwater lake. The gardens are formed by a series of terraces cut out of the hillside. Plants as diverse as palms, flame trees, cacti, pelargoniums, agapanthus, genistas and echiums all grow together here in abundance.

**Variation on the garden gnome**
A ship's figurehead from the Valhalla Museum is situated within the Abbey Gardens at Tresco and forms part of the National Maritime Museum.

**A show of colour**
Brightly coloured rock plants cling to old stonework.

**The pond garden**
This is one of the oldest parts of the garden. Originally there were three rectangular pools stocking carp for the kitchen.

**The arrival of spring**
Hampton Court at this time of year is a delight.

# Hampton Court
## PALACE GARDENS – BY ROYAL APPOINTMENT

Few other gardens have had the rapt attention of so many enthusiastic monarchs assisted by so many distinguished gardeners. Royalty has resided at Hampton Court from the Tudors to the House of Hanover, whilst gardeners have ranged from Henry Wise to 'Capability' Brown. Extensive restoration work has now been completed on Wise's privy garden including Tijou's magnificent wrought-iron screens.

Henry VIII's formative heraldic garden was not long-lasting, yet it was to pioneer gardens on a grand scale all over England, and most specifically at Hampton. It was Charles II, returning from exile in France and determined to match Vaux-le-Vicomte and Versailles, who conceived the spectacular Long Water. Although Charles was a keen gardener, few of his grand schemes were ever completed and it was William III and Mary in the late seventeenth century who set the scene for the gardens as they appear today. Hampton was to become their main residence and, after Whitehall Palace went up in flames, England's greatest architect, Sir Christopher Wren, was commissioned to build the east front that faces the vast semi-circular fountain garden.

Three hundred years later, Hampton Court is a marvellous retreat from city life, and thanks to Queen Victoria, not just for royalty but for everyone to enjoy.

## *S*YON PARK – A VERY GLAMOROUS GLASSHOUSE

The Duke of Northumberland's London address is a grand Tudor house with an interior designed by Robert Adam and a park created by 'Capability' Brown. Yet the real joy of Syon is undoubtedly Charles Fowler's enormously elegant Great Conservatory. Planted for scent and gardened organically, it provides a tranquil haven in an urban setting.

## *T*HE ROYAL BOTANIC GARDENS – A COLLECTOR'S PARADISE

An internationally renowned botanic institution occupying a 300-acre site to the west of London, Kew is worth a visit at any time of year. Whilst the pagoda and Decimus Burton's Palm House are famous landmarks, Queen Charlotte's cottage is a delight, particularly when the bluebells are out.

**The famous Hampton Court Maze**

Planted in 1690, this is the oldest hedge-planted puzzle maze in Britain.

**Imposing statue**
*(centre left)*

At one time there were thirteen fountains in the great fountain garden where this statue now stands. Unfortunately, none of them ever operated properly. The fountain garden was designed by the French man, Daniel Marot.

# *C*AMBRIDGE – THE OLDEST GARDEN CITY

The Backs are the gardens of some of Cambridge University's oldest colleges which back on to the river, creating a long green corridor through the heart of the city and bringing the country into the town. King's College paid a farmer to keep cows in the meadow opposite their chapel to enhance this rustic image.

The best way to appreciate the Backs is by river. Hire a punt from the Mill Pool, just above the bridge at Silver Street, find someone skilled at this ancient yet unwieldy form of transport, then relax and enjoy the view as you drift downstream. The gentle mile-long voyage down the slow-moving Cam to Magdalene Bridge passes some of the finest medieval architecture in Europe set amongst immaculate gardens – surely one of the world's greatest short journeys.

The Backs were the scene of one of 'Capability' Brown's rare setbacks. He had conceived a grandiose scheme to divert the river and combine all the college gardens into one park, but fortunately the dons rejected his plans.

**King's College Chapel**
Wide open lawns provide the river frontage to Cambridge's greatest medieval building.

## THE BIRMINGHAM BOTANICAL GARDENS AND GLASSHOUSES

Less than two miles from the city centre, these gardens are a true oasis, covering fifteen acres and containing a huge variety of features, ranging from a fine palm house to a show of carnivorous plants. The famous Victorian plant collector, E H Wilson, worked here in his youth and there is now a special display of the specimens he brought back from China. The gardens also house the National Bonsai Collection.

**Dahlias in Magdalene College gardens**

Magdalene College was the last of the colleges to admit female students.

**New Court, St John's College**

This part of St John's College is nicknamed 'the wedding cake' because of its romantic neo-Gothic architecture.

**A massive leaf**
The head gardener (1880) displaying a massive gunnera leaf in front of the pineapple pit and melon house.

# THE LOST GARDEN OF HELIGAN

Heligan owes its present glory to the endeavours of Tim Smit and John Nelson who, in the largest garden restoration project in Europe, have transformed a neglected wilderness into one of England's finest romantic gardens.

In the northern gardens are rockeries, summerhouses and a walled kitchen and flower garden, while to the south lies 'the jungle', a sub-tropical valley overlooking Mevagissey, which is rich in palms, bamboos, tree ferns, gunnera and other exotic trees and shrubs. Remarkably, much of the original plant collection has survived the seventy-year period of neglect, including many Hooker rhododendrons and early japonica camellias.

### New Zealand or Cornwall?

Giant ferns grow to massive proportions in the area known as New Zealand.

### Rhododendrons in both quality and quantity

The gardens at Heligan contain rhododendron specimens brought back from the Himalayas in the nineteenth century by Sir Joseph Hooker. They also contain the first kiwi fruit grown in England.

## PAINSWICK ROCOCO GARDEN – A SPRING GARDEN

Lord and Lady Dickinson, the owners of Painswick, have funded much of the recent restoration work on this rare rococo garden. Hidden away in a quiet corner of the Cotswolds, the nearby woodlands offer an extra bonus - one of the best snowdrop displays in England.

**Lime Avenue**

Biddulph is a hugely
varied garden with
shady walkways.

**Ready to leap**

This large stone frog is
one of the many exotic
creatures to be found at
Biddulph.

# BIDDULPH GRANGE
## AN ENDANGERED SPECIES

When the National Trust acquired Biddulph in 1988, the gardens were badly neglected. Indeed, there was a grave risk that this outstanding High Victorian garden, the creation of the industrialist James Bateman, might disappear completely. Today, skilful horticultural detective work, sheer physical labour and considerable expenditure have all contributed to the restoration of this fourteen-acre garden. Bateman had a passion for orchids, and the Trust has successfully recreated his remarkable blend of exotic planting and eclectic garden building.

Biddulph is a series of gardens within a garden, connected by tunnels and paths. Discover for yourself a pagoda, a stone sphinx, a huge gilded water buffalo and bizarre stone creatures as you wander through the grounds.

## The Chinese garden

In the Chinese garden at Biddulph, plants originally brought back from the Orient include the first golden larches, variegated bamboo, mountain peonies and Japanese anemones. There is even a willow-pattern-style Chinese bridge.

### Beware the golden buffalo

In a very secluded area of Biddulph, a newly gilded water buffalo surveys the scene.

## WESTBURY COURT
### A DUTCH INSPIRATION

Miraculously preserved, Westbury Court is one of the oldest surviving gardens in the country, providing a rare glimpse of a seventeenth-century Dutch water garden. The reflective qualities of its long, cool canals seem to contrive a feeling of both intimacy and spaciousness.

Few gardens survive the destruction of the house. Westbury Court saw the demolition of three houses before it was purchased by the National Trust in 1967. Today it is a shining example of a garden of the time of William of Orange, with almost all the specimens dating back prior to the eighteenth century. Ironically, in Holland, no such garden survives from this period.

# Westonbirt – a woodland garden on a grand scale

When contemplating the creators of great gardens, Captain Robert Holford does not readily spring to mind, yet here was a man who originated what many now consider to be the finest arboretum in the country.

Holford came from a very wealthy family and although he was greatly knowledgeable about horticulture, trees and shrubs were his first love. Like so many of the Victorian landowners, he commissioned plant hunters to bring back rare and unknown specimens from all over the temperate world, and was even rumoured to have imported soil from China! Nurseries were established throughout his estate with no effort spared in experiments with propagation and hybridisation. By the time he died towards the end of the nineteenth century, Westonbirt had acquired an international reputation.

Today the arboretum is expertly managed by the Forestry Commission. Spread over 500 acres, it contains 18,000 trees, of which some 4,000 are specimen trees, with a further 100 acres of downland.

**A good walk**

Broad avenues radiate out for miles in all directions; it takes about two hours to walk across the arboretum from end to end.

**Westonbirt ablaze**
It is in autumn that Westonbirt is particularly dramatic with different acers glowing in brilliant colours.

# LEVENS HALL – TOPIARY AT ITS BEST

Levens Hall is an Elizabethan house set in substantial parkland where black fallow deer and long-horned Bagot goats roam. Located on the southern edge of the Lake District, its origins date back 300 years, when Colonel James Grahme won the estate during a game of cards. When James II was forced to abdicate, the colonel retired to the north with Guillaume Beaumont, the ex-king's gardener. This unusual partnership produced a fascinating late-seventeenth century garden, whose highlight is the famous topiary.

With over ninety giant yews carved in wondrous shapes, it seems ironic that a gardener who had been forced to 'down tools' by William of Orange should subsequently fashion a garden in the Dutch style.

**Bizarre shapes**
The topiary at Levens Hall is very old. Some trees were planted more than 300 years ago and carved in the most unusual shapes, such as chess pieces, lions, peacocks and great umbrellas. There have been less than a dozen head gardeners employed in the three centuries of Levens Hall's existence.

South II 1992 by Mimmo Paladino (above)

*Three Piece Reclining Figure Number One* by Henry Moore

# THE YORKSHIRE SCULPTURE PARK
## A HAVEN FOR CONTEMPORARY SCULPTORS

Both of Britain's greatest twentieth-century sculptors, Henry Moore and Barbara Hepworth, were born in Yorkshire: Moore at Castleford, and Hepworth at nearby Wakefield. So it is therefore highly appropriate that the nation's largest sculpture park should be located in Yorkshire, close to both Castleford and Wakefield.

Occupying one hundred acres of eighteenth-century parkland on the former estate of Bretton Hall, the Yorkshire Sculpture Park was established in 1977. A major exhibition in May 1994, entitled

'Henry Moore in Bretton Park', and displayed in the neighbouring deer park, established a permanent exhibition of more than a dozen of his works.

The Sculpture Park is one of the few outdoor art galleries in Europe which is totally devoted to twentieth-century sculpture. Through a series of ever-changing exhibitions, the works of such pre-eminent British artists as Elisabeth Frink, Anthony Caro, Bernard Meadows and Barbara Hepworth are displayed, as well as those of overseas artists such as Max Ernst and Kan Yasuda. The Yorkshire Sculpture Park also gives support to young artists through bursaries and residencies, arranges workshops, and has designed a five-acre Access Sculpture Trail to cater for disabled visitors.

**The world of Kan Yasuda** (*above*)

'In the real three-dimensional world in which we live, I search for another dimension where the space allows greater, or some other kind of understanding that comes from knowing within the heart' (Kan Yasuda).

**Thsuki-no-hikari** (*above*)

The lines of this dramatic form by Igor Mitoraj are thrown into sharp relief in winter, when the branches are bare and snow lies on the ground.

**The Access Sculpture Trail**

This is an innovative yet gentle introduction to sculpture with particular emphasis on the needs of disabled visitors. The British sculptor, Don Rankin, was commissioned to create the trail in 1985, and the result is designed to stimulate the senses with a variety of textures, smells, colours and sounds.

**Ithaka** (*right*)

This bronze, also by Igor Mitoraj, was created in 1991 and exhibited at the Yorkshire Sculpture Park during the autumn and winter of 1992/93.

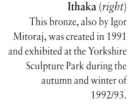

**In Memoriam**

In a sunlit clearing close to the entrance is Elisabeth Frink's meditative 'In Memoriam', a bronze sculpture that she created in 1983.

# ℛOCHE COURT
## A GARDEN OF METAL AND STONE

This secluded estate near Salisbury is not in itself a great garden, according to Lady Bessborough, yet she has made it so in a fascinating way, planting sculptures on the lawns amongst the trees and on the parkland. Having owned a London gallery for nearly forty years, this seemed a natural progression, and she began some seven years ago with an exhibition of five sculptures, which included works by Barbara Hepworth and Elisabeth Frink.

Even gardens devoted to sculpture grow and now there are over one hundred pieces representing the cream of contemporary sculpture. There are sculptures everywhere: Tim Harrisson's huge wood carving 'Heaven's Gate' at the entrance; Barbara Hepworth's 'Pierced Monolith with Colour' on the lawn in front of the house; American Jim Dine's bronze 'Little Lady' stands outside the Orangery; whilst Frink's 'In Memoriam' looks contemplative in a wooded glade. To complete the scene, enormous terracotta pots loom in long grass amongst the terracotta-coloured Limousin cattle.

**Night Gesture**

Constructed of wood and aluminium, this dramatic piece by Louise Nevelson stands at approximately three metres high.

**The Garden of Epicurus** *(left)*

These four copper heads by Peter Burke gaze impassively into space at the far end of the garden.

## THE BARBARA HEPWORTH SCULPTURE GARDEN
### THE ARTIST'S GARDEN

Hepworth laid out the garden herself, stipulating that the sculptures should remain a permanent feature. Positioned between trees, bushes and

**Tim Harrisson's Heaven's Gate**

This was carved from the wood of a douglas fir imported from the Longleat estate. Originally dark-coloured, it has weathered over the years to a soft pale grey.

**Figure Lying on its Side**

This figure, by Ken Armitage, is watched by Reg Butler's 'Girl', partially hidden in the bushes at Roche Court.

semi-tropical flowers, it is an enchanting, highly unusual garden, looking out over the rooftops of St Ives to the Atlantic Ocean.

1. **Barbara Hepworth Sculpture Garden**
   Barnoon Hill, St Ives, Cornwall
   TR26 1TG  *(01736) 796226*

2. **Barrington Court,** Barrington,
   near Ilminster, Somerset TA19 0NQ
   *(01460) 241938*

3. **Biddulph Grange Garden,** Biddulph
   Stoke-on-Trent, Staffordshire
   ST8 7SD  *(01782) 517999*

4. **Birmingham Botanical Gardens**
   Westbourne Road, Edgbaston
   Birmingham B15 3TR
   *(0121) 454 1860*

5. **Blenheim Palace,** Woodstock
   Oxfordshire OX20 1PX
   *(01993) 811091*

6. **Bowood House,** Calne, Wiltshire
   SN11 0LZ  *(01249) 812102*

7. **Broadlands,** Romsey, Hampshire
   SO51 9ZD  *(01794) 516878*

8. **Buscot Park,** Faringdon, Oxfordshire
   SN7 8BU  *(01367) 242094*

9. **Cambridge,** the Backs

10. **Castle Howard,** York, N Yorkshire
    YO6 7DA  *(01653) 648444*

11. **Chatsworth,** Bakewell, Derbyshire
    DE45 1PP  *(01246) 582204*

12. **Coleton Fishacre Garden**
    Coleton, Kingswear, Dartmouth,
    Devon TQ6 0EQ  *(01803) 752466*

13. **Hampton Court Palace,** East Molesey
    Surrey KT8 9AU  *(0181) 781 9500*

14. **Harewood House,** Harewood, Leeds
    W Yorkshire LS17 9LQ
    *(0113) 288 6331*

15. **Heligan,** Pentewan, St Austell
    Cornwall PL26 6EN  *(01726) 844157*

16. **Hestercombe,** Cheddon Fitzpaine,
    near Taunton, Somerset TA2 8LQ
    *(01823) 337222*

17. **Hyde Hall Garden,** Rettendon
    Chelmsford, Essex CM3 8ET
    *(01245) 400256*

18. **Levens Hall,** Kendal, Cumbria
    LA8 0PD  *(015395) 60321*

19. **Montacute House,** Montacute
    Somerset TA15 6XP  *(01935) 823289*

20. **Mottisfont Abbey Garden**
    near Romsey, Hampshire SO51 0LP
    *(01794) 340757*

21. **Mount Edgcumbe,** Cremyll
    near Torpoint, Cornwall PL10 1HZ
    *(01752) 822236*

22. **Overbecks Garden,** Sharpitor
    Salcombe, Devon TQ8 8LW
    *(01548) 842893*

23. **Painswick Rococo Garden,** Painswick
    Gloucestershire GL6 6TH
    *(01452) 813204*

24. **Roche Court Sculpture Park**
    East Winterslow, Salisbury, Wiltshire
    SP5 1BG  *(01980) 862244*

25. **Royal Botanic Gardens,** Kew
    Richmond, Surrey TW9 3AB
    *(0181) 940 1171*

26. **St Michael's Mount,** Marazion
    near Penzance, Cornwall TR17 0HT
    *(01736) 710507*

27. **Sheffield Park,** Uckfield, E Sussex
    TN22 3QX  *(01825) 790231*

28. **Sissinghurst Castle Garden**
    Sissinghurst, near Cranbrook, Kent
    TN17 2AB  *(01580) 712850*

29. **Stourhead,** Stourton, Warminster
    Wiltshire BA12 6QD  *(01747) 841152*

30. **Studley Royal Water Garden**
    Fountains, Ripon, N Yorkshire
    HG4 3DY  *(01765) 608888*

31. **Syon Park,** Brentford, Middlesex
    TW8 8JF  *(0181) 560 0881*

32. **Tresco Abbey Gardens,** Tresco, Isles
    of Scilly TR24 0QQ  *(01720) 422849*

33. **Westbury Court Garden**
    Westbury-on-Severn
    Gloucestershire GL14 1PD
    *(01452) 760461*

34. **Westonbirt Arboretum**
    Tetbury, Gloucestershire
    GL8 8QS  *(01666) 880220*

35. **Yorkshire Sculpture Park**
    Bretton Hall
    West Bretton, Wakefield
    W Yorkshire WF4 4LG
    *(01924) 830302*

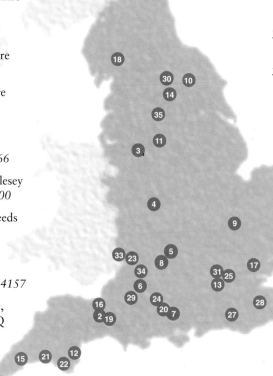

*Note: whilst all gardens are open
to the public, opening hours vary
considerably. It is therefore advisable
to check before visiting.*